To my rock star sister, Lynda, the best friend to explore rocks with growing up
—S. N. W.

For Debra
—N. C.

Simon & Schuster Books for Young Readers • An imprint of Simon & Schuster Children's Publishing Division • 1230 Avenue of the Americas, New York, New York 10020 • Text © 2025 by Sandra Neil Wallace • Illustration © 2025 by Nancy Carpenter • Back matter background by paladin13/iStock • Book design by Sarah Creech • All rights reserved, including the right of reproduction in whole or in part in any form. • SIMON & SCHUSTER BOOKS FOR YOUNG READERS and related marks are trademarks of Simon & Schuster, LLC. • For information about special discounts for bulk purchases, please contact Simon & Schuster Special Sales at 1-866-506-1949 or business@simonandschuster.com. • The Simon & Schuster Speakers Bureau can bring authors to your live event. For more information or to book an event, contact the Simon & Schuster Speakers Bureau at 1-866-248-3049 or visit our website at www.simonspeakers.com. • The text for this book was set in Neighbor and Fantabular MVB. • The illustrations for this book were rendered with a combination of acrylic paint and digital tools. • Manufactured in China • 0625 SCP • First Edition
2 4 6 8 10 9 7 5 3 1
Library of Congress Cataloging-in-Publication Data
Names: Wallace, Sandra Neil, author. | Carpenter, Nancy, illustrator.
Title: Rock star : how Ursula Marvin mapped moon rocks and meteorites / Sandra Neil Wallace ; illustrated by Nancy Carpenter.
Other titles: How Ursula Marvin mapped moon rocks and meteorites
Description: First edition. | New York : Simon & Schuster Books for Young Readers, [2025] | "A Paula Wiseman Book." | Includes bibliographical references. | Audience: Ages 4–8 | Audience: Grades 2–3 | Summary: "A biography of Ursula Marvin, who spent 50 years breaking barriers in science and becoming a pioneer among women reshaping American science"—Provided by publisher.
Identifiers: LCCN 2024044903 (print) | LCCN 2024044904 (ebook) | ISBN 9781534493339 (hardcover) | ISBN 9781534493346 (ebook)
Subjects: LCSH: Marvin, Ursula B.—Juvenile literature. | Women geologists—United States—Biography—Juvenile literature. | Geologists—United States—Biography—Juvenile literature. | Lunar petrology—Juvenile literature. | Meteorites—Antarctica—Juvenile literature.
Classification: LCC QE22.M36 W35 2025 (print) | LCC QE22.M36 (ebook) | DDC 551.092 [B]—dc23/eng/20241127
LC record available at https://lccn.loc.gov/2024044903
LC ebook record available at https://lccn.loc.gov/2024044904

ROCK STAR

How Ursula Marvin Mapped Moon Rocks and Meteorites

Written by
Sandra Neil Wallace

Illustrated by
Nancy Carpenter

A PAULA WISEMAN BOOK
Simon & Schuster Books for Young Readers
New York   Amsterdam/Antwerp   London
Toronto   Sydney/Melbourne   New Delhi

Ursula grew up in a small Vermont village with big mountains to explore.

They wrapped around her house like a crescent moon.

With Coolidge the cat (named after the president) and Nita the Saint Bernard, Ursula hiked the green peaks until the sky glowed fiery orange.

Winter brought the best adventures. Howling snowstorms. Potato fields covered in ice. Ursula strapped on her skis and roared across them, pulled by the neighbor's horse.

The cold air felt delicious. Ice hummed beneath her skis. But what Ursula loved most was how beautiful the mountains looked under the frosty moonlight.

When ice fields became vegetable fields again, Ursula collected potato beetles with her father. As Vermont's official bug doctor, he studied insects devouring crops.

Ursula had no intention of becoming an entomologist or any kind of scientist.

She was meant to explore.

But in college, examining a rock under a microscope lit a fire in her. Suddenly, Ursula had to be a scientist and find out what the mountains she loved were made of.

Ursula's professor disagreed. He blocked her from majoring in geology because she was a woman. He told Ursula she should be learning to cook.

Ursula ignored him. She kept examining rocks at a different school and became a geologist.

Ursula was so good at recognizing rare minerals in Earth rocks that she became one of the first geologists to study the rocks Apollo astronauts brought back from the moon.

Squeezing her arms into a glove box, Ursula cradled the moon rocks, searching for the unusual. She gazed into a microscope to spot something rare. What she found surprised her. The rocks sparkled with speckles of glass and minerals. Shiny black and cloudy white. Swirling yellow and fiery orange. Bursting with color, like a stained-glass window.

Ursula uncovered minerals on the moon that showed its surface was once a bubbling ocean of melted rock. This new theory replaced the old science about the moon.

Ursula examined more alien rocks—meteorites—that catapulted from space and landed on Earth in fiery explosions. X-raying their crusty coatings, she spotted minerals no one knew existed beyond Earth. Ursula revolutionized how scientists saw the solar system.

But like the astronauts on the moon, Ursula was meant to explore. She yearned for adventure and to find meteorites in Antarctica. Its mountains held the rarest specimens blasted from asteroids and maybe even the moon.

No woman had searched for meteorites at the bottom of the world. Ursula wanted to be the first.

Ursula climbed into the belly of the plane that would take her deeper into Antarctica. She rocked to the roar of the engine and eyed the cargo in front of her: ice axes, snow shovels, boxes for meteorites marked DEEP FREEZE, her pink duffel bag crammed with fifty pounds of clothing. (She had to keep warm in the coldest place on Earth!) Ursula had stuffed the bag with thirty-four mittens, twenty-four socks, eighteen turtlenecks, six wind pants, and a thick, downy robe for trips to the toilet tent.

With a ROAR and a RUMBLE, the plane took off.

Below, a smoking volcano spewed lava onto the ice.
A family of penguins slid into the Southern Ocean.

At Ursula's camp, there were no penguins. No erupting volcanoes.

Far from the sea, the wind blew harder. The temperature plunged lower.

Ice gathered around the Transantarctic Mountains, carrying frozen treasure: meteorites, perfectly preserved. Ursula and her teammates didn't have much time to discover them before Antarctica's winter storms buried the meteorites for another year.

They set up camp on the oldest ice in the world: hoisting tents, filling stoves, unrolling sleeping bags. Battening down every object the wind would blow away. Nothing would be left behind.

With the mountains wrapped around their camp like a crescent moon, Ursula and scientists from around the world shared pastrami sandwiches and Japanese noodles.

Ursula was the only woman.

At night, the sun shone like it was day, streaming lemon yellow into Ursula's tent. She couldn't sleep. And she felt nervous.

Would she find a meteorite?

She had studied meteorites under a microscope. But could she spot them on acres of ice? Could she sort them from ordinary Earth rocks?

In the morning, Ursula pulled on her parka. She laced up her snow boots.

But she felt like a clown.

"Parka too big—wind pants too big—boots too big." Ursula sighed. How would she find anything in clumsy clothing meant for men?

Ursula's snowmobile was the perfect size. She named it *Blue Ice*.

Riding *Blue Ice* through a moraine piled with rocks, Ursula spotted something in the snow. A lumpy, brown rock. Could it be a meteorite? She threw off her mittens, took out her hand lens, and examined it.

This rock had a crust! A shiny layer that melted on its way to Earth. Ursula had found her first meteorite!

She stuck a red flag in the ice. Using sterilized tongs, Ursula placed the meteorite in a Teflon bag, then into the icebox on her snowmobile.

The next time she spotted a meteorite, when she reached for it, the meteorite was gone! Tossed away by blowing snow with a windchill of forty below.

Some teammates shivered. Not Ursula. She found the icy air delicious. She kept riding *Blue Ice* and collecting meteorites more than four billion years old!

Wavy, black meteorites on ice fields, turquoise blue.
Specimens as small as berries on rippled gray slopes.
Foot-long meteorites older than Earth, their crusts a velvety brown like the skua birds flying above, and just as rare.
Whenever Ursula discovered a meteorite, she heard the ice hum beneath her snowmobile.
That week, the ice hummed 159 times.

"It is beautiful here," Ursula declared. "And a truly great adventure."

One day, a teammate broke the ski on his snowmobile. Ursula shared her ride.

But Ursula had never climbed a nunatak before. As high as a mountain, the rocky ridges loomed large in Antarctica.

"We are hiking to the top," her teammate announced. Ursula wasn't sure she could.

She looked down at her too-big boots and began to climb. She was the last one to reach the top, but she made it.

Still, Ursula began to doubt. Did she really belong in Antarctica?

Flagging a beautiful specimen, Ursula looked closer and saw that it wasn't extraterrestrial, but an ordinary Earth rock. Maybe she wasn't so great at finding meteorites after all.

On the ride back to camp, the ice no longer hummed. That night, thinking about the day, Ursula burned her steak dinner.

She listened to the growing wind roll in from the South Pole. A storm was coming. A powerful force called a katabatic wind blew across the ice caps and roared into Ursula's tent. Lifting the door flap, Ursula was pelted with snow. She couldn't see past the toilet tent. She pulled her sleeping bag over her face and tried to sleep.

The next morning, Ursula woke up early. She ate oatmeal for the nineteenth day in a row.

But she was told to stay back. Too many people. Not enough snowmobiles. Furious, Ursula protested, but it didn't work.

Alone in the camp, she melted ice for drinking.

She played with her Rubik's cube.

Nothing helped Ursula feel better. She was meant to explore.

When her teammates returned, Ursula hopped on her snowmobile and sped off to the ice fields.

She needed to find a meteorite.

She spotted one in a wind scoop carved into the ice. That made her feel better. Carrying it home, Ursula vowed she would never let anyone stop her from exploring again.

With one week left to search for meteorites, Ursula expertly zigzagged between dangerous cracks in the ice to unexplored places, where the meteorites were bigger and rarer. She kept finding more treasure and planned to discover the last meteorite of the season.

Number 300, Ursula predicted. Would it be a piece of the moon?

As the ice hummed beneath her snowmobile, Ursula started to sing. Happy and free, she finally knew she could do anything. Belong anywhere. Especially in Antarctica.

She could find meteorites. She had the strength to be an explorer, and she thrived in the coldest place on Earth.

With her confidence soaring, Ursula rode her mighty machine across the snowy desert, dreaming beyond science. In her imagination, she was a snowmobile racer, competing with style and spirit.

Faster and faster Ursula rode, carving her own path under the midnight sun until she reached her yellow tent.

Back at camp, Ursula threw a cover over her snowmobile.
A gust of wind blew the cover into the sky.
Ursula reached for it. She crashed to the ice.
Something bad just happened, Ursula thought.
She'd twisted her leg, but was it broken?

Ursula crawled into her tent.
"It might not be as bad as I thought," she told her teammates.
But Ursula worried. She couldn't sleep. She needed to find the last meteorite.
How could she do that with an injured leg?

The next day, Ursula tried a shovel as a crutch. She still couldn't walk. She had to be airlifted to Antarctica's hospital.

But not without her specimens.

Waiting to be rescued, Ursula sat on the icebox filled with her meteorites.

Strapped inside the helicopter, Ursula rocked to the roar of the engine.

She reached for her leg—purple, puffy, and painful.

She looked out at the bumpy blue ice that she loved, inching farther away.

"This is the end of my great adventure." She sighed.

When the plane landed, nurses rushed Ursula to the hospital. But they didn't expect a woman explorer and handed her men's pajamas.

"You've got a broken leg," the doctor said.

"You are tough," the nurses agreed. They covered Ursula's leg with so many ice packs, it looked like a nunatak.

Lying in the hospital bed, Ursula thought of her teammates one hundred miles away. "I am very sorry I have missed this last week," she wrote.

As snow blew across the Transantarctic Mountains, Ursula's teammates collected the final meteorite.

Number 373—a season record.

This meteorite looked different. Where was it from? An asteroid? Or . . . could it possibly be . . . a piece of the moon?

Back at home, Ursula couldn't wait to examine it.
Squeezing her hands into a glove box, Ursula cradled the meteorite.
The size of a plum, it glinted, bubbly green and tan.
It sparkled with speckles of dots.
Peering into the microscope, Ursula saw that the dots were minerals and glass. Shiny black and cloudy white. Swirling yellow and fiery orange. Bursting with color, like a stained-glass window.
Just like the rocks the astronauts brought back from the moon!

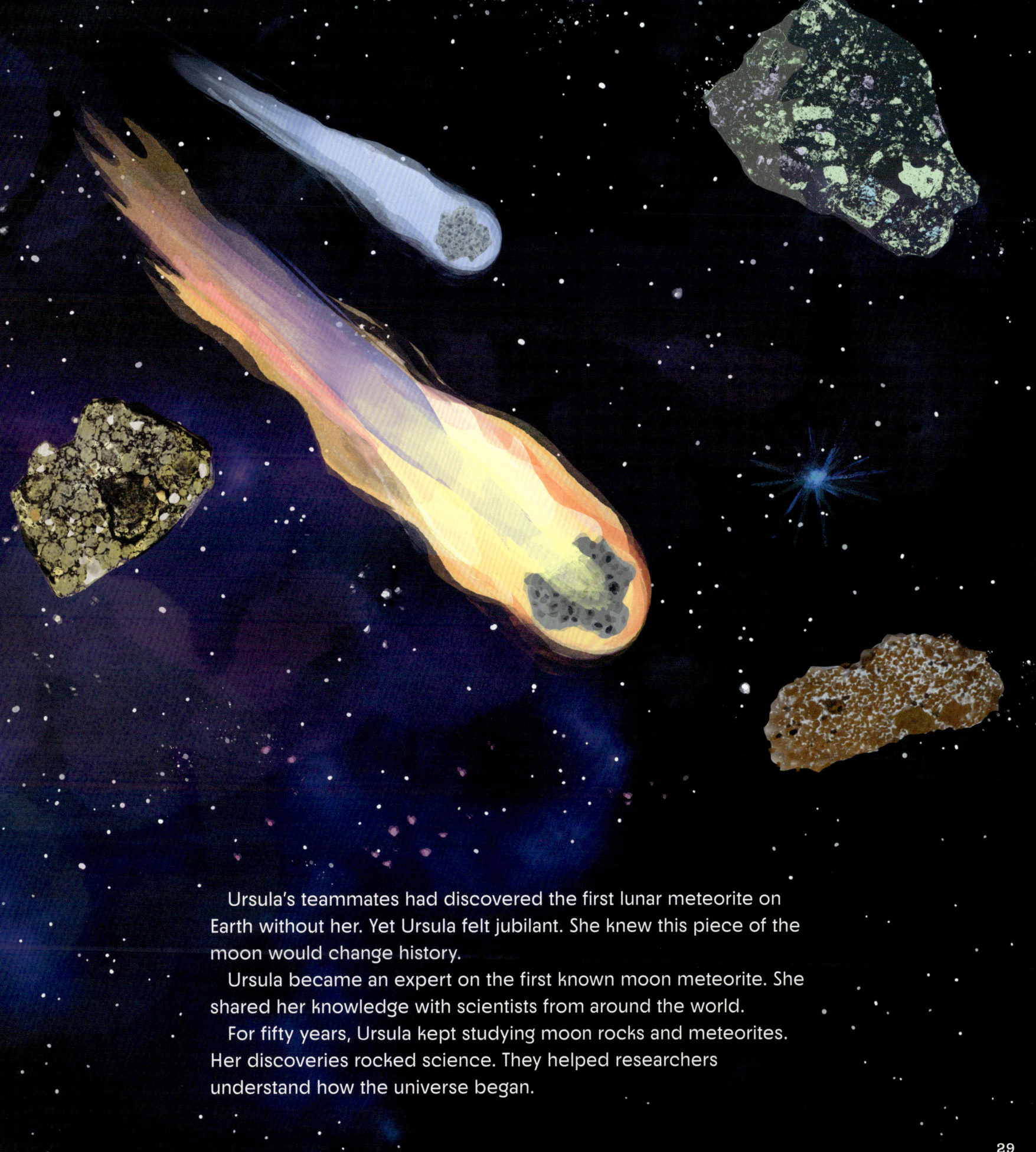

Ursula's teammates had discovered the first lunar meteorite on Earth without her. Yet Ursula felt jubilant. She knew this piece of the moon would change history.

Ursula became an expert on the first known moon meteorite. She shared her knowledge with scientists from around the world.

For fifty years, Ursula kept studying moon rocks and meteorites. Her discoveries rocked science. They helped researchers understand how the universe began.

Because of Ursula, more women became scientists. Exploring Antarctica on their snowmobiles, they found more moon meteorites.

They discovered Martian meteorites four billion years old!

Near their discoveries is a rocky ridge, jutting into the sky seven thousand feet high, known as Marvin Nunatak. Far above it, between Mars and Jupiter, Asteroid Marvin orbits the Sun. On the rocky lunar surface, a piece of the moon is called Marvin Crater. All are named after Ursula Marvin—rock star of the universe.

AUTHOR'S NOTE

*"I really would not exchange for anything . . . the thrill
of seeing those first samples from the moon or spotting
black rocks on the Antarctic ice."* —Ursula Marvin

In 1978, Ursula Bailey Marvin became the first woman to search for meteorites in Antarctica. She traversed ancient ice no human had been to and found rare meteorites new to science.

As a visionary and groundbreaking geologist, Ursula also had to be brave. She faced fierce winds, the coldest climate on Earth, and ice that could have crushed her. But Ursula believed that meteorites—along with moon rocks—held answers to unlocking the origins of the solar system. For her, facing the danger was worth the risk.

Ursula also faced gender discrimination. Like the early women astronauts, she was issued expedition clothing in men's sizes. The US Navy and the National Space Foundation prevented women from being in Antarctica and conducting scientific research there until 1969. But Ursula knew that science exploration wasn't just for men. She shattered this myth by trekking to Antarctica three times between 1978 and 1985—twice to search for meteorites and once to find evidence on why dinosaurs became extinct.

Ursula's extraordinary meteorite expeditions and the journals she wrote—describing her fears and discoveries; what she imagined, packed, and ate; and the agony of breaking her leg—form the focus of this book. I've tried to stay in the mindset of what scientists knew about moon rocks and meteorites during Ursula's expeditions. For instance, scientists didn't yet know for certain that meteorites could be pieces from a planet or the moon until the team Ursula was part of discovered the first known lunar meteorite on Earth in 1982.

Ursula was born in Bradford, Vermont, in 1921 and grew up exploring nature with her brother and sister and the family pets. She went skijoring over ice, pulled by a horse. But Ursula never imagined becoming a scientist until taking a geology course at Tufts University in the 1940s.

Back then, it was nearly impossible for women to have science careers. Her male professor had the power to block her from majoring in geology, and he did. Furious yet determined, Ursula earned her geology degrees at Radcliffe College and Harvard University and spent the next fifty years breaking barriers in science.

She became the first woman faculty member in Harvard's geology department and the first woman president of the Meteoritical Society. She pushed for equal access and pay equity for women scientists by launching the first women's program at the Smithsonian Astrophysical Observatory—the location of Ursula's main laboratory.

During those years, Dr. Marvin shaped planetary science. She participated in major twentieth-century discoveries and developed science theories that are now foundational.

Ursula was the first person to x-ray meteorites for rare minerals (no one had thought of it before).

She changed how scientists see the solar system. Studying the Allende meteorite that smashed to Earth in 1969, Ursula uncovered minerals in this early rock that contained the building blocks of life. Her groundbreaking scientific article on Allende created the study of CAIs (calcium–aluminum-rich inclusions).

Ursula helped uncover the Lunar Magma Ocean theory. As one of the first geologists recruited by NASA to study lunar rocks the Apollo astronauts brought back from the moon, Ursula examined samples from every Apollo mission and the Soviet Union's (now Russia) Luna robotic probe samples too. She discovered minerals that showed the early moon's surface was once a magma ocean. This revolutionized what scientists knew about the moon's beginnings.

And then, of course, Ursula became an expert on the first known moon meteorite.

But searching for meteorites in Antarctica was what Ursula cherished the most. Not because scientists discovered, decades later, that some of Ursula's rare meteorites had catapulted from Vesta, the brightest asteroid in the sky. Or because she'd made history by breaking a gender barrier and being part of the team that discovered the first known moon meteorite on Earth.

It was the thrill of exploring the frozen slopes with her snowmobile, *Blue Ice*, and spotting an extraterrestrial rock in the snow.

"Riding through the vastness, the glimpse of a dark object starts the heart pounding," Ursula wrote. "Racing toward it the excitement grows as one sees it is not a shadow, not a glacial cobble, but a meteorite—a piece of rock from another planet!"

Ursula Marvin, truly a rock star of the universe, died in 2018 at the age of ninety-six.

Ursula in Antarctica.

IN HER OWN WORDS
From Ursula's Antarctica Journals

December 12, 1978

"Air clean, scenery across ice to Transantarctic range magnificent. But best of all, Mt. Erebus, puffing away."

December 29, 1978

"I have never seen such wind. It sailed over slopes with wisps of snow straight in our faces."

January 5, 1979

"As we rode I found myself singing in the wondrous, star-sown night."

January 8, 1979

"Snowmobiling on the high blue ice with drifting snow was great fun, although I kept having to stop and crank mine up, and once to pull it upright when it overturned."

January 6, 1982

"There are many more acres of ice in the Allan Hills, and I am more enthusiastic than before about getting at it. Finally I see that I will hold out—can do it, enjoy doing it, am holding my own."

January 10, 1982

"Wind kept up all night. It was fun to listen to it. Powerful gusts, then all quiet."

ANTARCTICA FACTS

1. Antarctica is the coldest, windiest, and driest place on Earth. Even though it's so cold, it's considered a desert, because it gets very little rain or snow. The snow that does fall becomes part of the colossal ice sheets covering most of the continent.

2. As the fifth-largest continent, Antarctica spans 5.4 million square miles—about the size of the United States and Mexico combined. Less than five thousand people live there at any given time. That's because a 1959 international treaty made Antarctica a place devoted to peaceful, scientific research. The US McMurdo Station is the most populous Antarctic science research station.

3. More than sixty percent of the world's meteorites have been discovered in Antarctica, including the first moon meteorite found by Ursula's expedition team in 1982. These extraterrestrial rocks are easier to spot on the ice, and the desert climate helps preserve them.

4. In 1978, Ursula Marvin became the first woman to search for meteorites in Antarctica as part of the American-led Antarctic Search for Meteorites Program (ANSMET). Because of Ursula, women scientists have been explorers on all but four of ANSMET's yearly expeditions since then.

5. Antarctica holds seventy percent of the Earth's water and ninety percent of its ice. The blue ice along the Allan Hills where Ursula searched for meteorites is the oldest ice in the world, dating back six million years.

6. Climate change is accelerating the rate of ice melt in Antarctica. Scientists are studying the ancient blue ice that Ursula trekked across and the carbon dioxide air bubbles inside it for clues about the planet's climate history.

7. Antarctica has two seasons: winter and summer. In the winter, Antarctica is on the side of the Earth

tilted away from the sun and experiences darkness for six months. In the summer, when scientists like Ursula recover meteorites, they are surrounded by twenty-four hours of daylight as the sun never sets, though the temperature rarely gets above freezing.

URSULA B. MARVIN AND HER MOON ROCK AND METEORITE MILESTONES

1921-August 20: Ursula Bailey is born in Bradford, Vermont, the third child of schoolteacher Alice Bartlett Bailey and entomologist Harold Bailey. The family moves to Montpelier, Vermont, when Ursula is eight.

1935: She attends Montpelier High School and graduates in 1939.

1940: Ursula majors in history at Tufts University but requests to change her major to geology. Professor Robert Nichols denies it, saying geology is no field for women.

1943: She graduates from Tufts University.

1943: Ursula attends Radcliffe College Graduate School of Arts and Sciences at Harvard University. It's the first time classes are coed, but the schools are still separated by gender.

1946: Ursula earns her master's degree in geology, specializing in mineralogy. She becomes the first woman research and teaching assistant in geology at Harvard and the first woman member of Harvard's Geology Club.

1950: Ursula is hired to examine the mineralogy of meteorites in Harvard's museum collection.

1952: She marries mining geologist Thomas Marvin and travels with him around the world to search for iron.

1958: Ursula accepts a teaching position at Tufts from the professor who blocked her from being a geologist.

1961: Ursula begins research at SAO—the Smithsonian Astrophysical Observatory (now the Center for Astrophysics | Harvard & Smithsonian). She will make groundbreaking discoveries for the next four decades.

1962: A chunk of the Soviet Union's satellite Sputnik 4 crashes to Earth. Ursula x-rays it and finds the rare mineral wüstite, not thought to survive in nature. She then becomes the first person to x-ray meteorites, also discovering wüstite, changing what scientists know about objects reentering Earth's orbit and the minerals in meteorites.

1969: Ursula earns her PhD in geology from Harvard University. She is now Dr. Marvin.

1969: The US Navy and the National Space Foundation finally end their bans preventing women from being in Antarctica and conducting scientific research there.

1969-February 8: Ursula examines the Allende meteorite that explodes over Pueblito de Allende, Mexico. She finds calcium–aluminum-rich inclusions, launching the study of CAIs and revolutionizing how scientists see the solar system.

1969-July 20: The first humans land on the moon. Apollo 11 astronaut Neil Armstrong collects lunar samples as Ursula waits on Earth to examine them. An original Apollo 11 sample coinvestigator, she uncovers lunar minerals that change what scientists know about the moon's beginnings with the Lunar Magma Ocean theory.

1969-November 19: Apollo 12 lands on the moon. Ursula studies its specimens. She will go on to examine samples from all Apollo missions.

1969-December 21: Japanese scientists discover meteorites from different falls in pristine condition in Antarctica, sparking interest in meteorite expeditions to the frozen continent.

1971-February 5: Apollo 14 lands on the moon.

1971-July 30: Apollo 15 lands on the moon. Ursula identifies a new type of lunar rock blasted from its deep crust, showing that the moon's craters are caused by impact events.

1972-February 21: The Soviet Union's Luna 20 collects moon samples, which Ursula studies.

1972-April 20: Apollo 16 lands on the moon. At NASA, Ursula examines and classifies its specimens through a nitrogen glove box.

1972-December 11: The last Apollo mission lands on the moon. Ursula is a member of the Apollo 17 Sample Preliminary Examination Team. She will continue to study moon rocks until 1995.

1973: Ursula publishes the book *Continental Drift: The Evolution of a Concept*, about the Earth's continents moving horizontally instead of staying fixed. She will go on to publish more than 160 scientific articles.

1974: As the first Federal Women's Program coordinator at the Smithsonian Astrophysical Observatory, Ursula pushes for gender equity in science.

1974: Ursula becomes the first woman faculty member of geology at Harvard.

1975: She creates and organizes the first Space for Women Conference.

1975: Ursula is named the Meteoritical Society's first woman president and becomes a meteorite historian, interviewing scientists about their discoveries.

1978-1979: Ursula becomes the first woman to search for meteorites in Antarctica. She is joined by a team of American and Japanese scientists. The expedition to the Transantarctic Mountains is part of the US Antarctic Search for Meteorites Program (ANSMET) and sponsored by the National Science Foundation.

1981-1982: Ursula returns to Antarctica to search for meteorites. Her team uncovers the first known moon meteorite—ALH81005—proving that pieces from the moon and planets travel across space and land on Earth. Before the expedition ends, Ursula breaks her leg. She is airlifted to the US Navy hospital at McMurdo Station.

Ursula tries on her men's parka before her 1978 Antarctica expedition.

1985: Ursula travels to Antarctica's Seymour Island to search for evidence of why dinosaurs disappeared. The samples in her specimen boxes are stolen.

1986: The Geological Society of America awards Ursula the History of Geology Award.

1991: Asteroid 4309 is named Asteroid Marvin.

1992: An Antarctic ridge is named Marvin Nunatak by the Advisory Committee on Antarctic Names.

1993: Ursula heads the Meteorite Working Group, which sends Antarctic meteorite samples to scientists worldwide.

1997: Ursula is awarded the Women in Science and Engineering (WISE) Lifetime Achievement Award.

2011: After the Dawn spacecraft explores Asteroid Vesta, scientists learn that some of the meteorites Ursula discovered are from Vesta.

2012: Ursula receives the Meteoritical Society Service Award for her research on the history of geology.

2018–February 12: Ursula dies in Concord, Massachusetts, at age ninety-six.

2021: A crater near the moon's south pole is named Marvin Crater, a possible landing site for the Artemis III moon landing.

ACKNOWLEDGMENTS

To tell Ursula Marvin's story, I needed access to archives that housed her collections or held interviews and articles not digitized. But in 2020, the Covid-19 pandemic forced these institutions to close to in-person visits. I had to become an intrepid reporter, conducting over fifty interviews by phone and virtually, as well as nearly one hundred email conversations. I am so grateful to the archivists, scientists, and family members of Ursula's who generously gave me the information I needed, including:

Gayl Heinz and Harold L. Bailey, who provided insight into their aunt Ursula's childhood and photos not found in any collection.

The Smithsonian's supervisory archivist, Tammy Peters, who was the Smithsonian Institution's chief archivist during the pandemic and who acquired the Ursula B. Marvin Collection. She worked tirelessly to provide me with primary print sources and images that were essential to telling this story. The Smithsonian's Astrophysical Observatory librarian, Maria McEachern, also dug deep to send me crucial articles about Ursula.

Groundbreaking scientist David Kring, who located the dinosaur-killing Chicxulub impact site in Yucatán, Mexico, and who worked with Ursula. Dr. Kring is the principal scientist at the Lunar and Planetary Institute and principal investigator of the Center for Lunar Science and Exploration. He studies the landing sites for NASA's moon missions, including the Artemis moon mission. Dr. King reviewed this manuscript for scientific accuracy.

Thanks to mountaineer John Schutt for his interviews and photos. John was Ursula's teammate on her second Antarctic expedition, and he spotted the moon meteorite. Thanks also to ANSMET's coleader James Karner and NASA Astromaterials Research and Exploration scientist David Mittlefehldt, who confirmed ages and parent bodies of Antarctic meteorites uncovered on Ursula's expeditions; and to Akira Yamaguchi of the National Institute of Polar Research, for verifying information about Japan's Antarctic research expeditions and explorers.

And to women rock star scientists Roberta "Robbie" Score—former NASA Antarctic explorer and NASA Antarctic Meteorite Lab manager, who worked with Ursula and discovered the famous Martian meteorite in 1984—and Barbara Cohen of NASA Goddard Space Flight Center, for answering all my questions and continuing to inspire kids to be scientist explorers.

QUOTE SOURCES

Page 12: **"Parka too big . . ."**: Ursula B. Marvin, *Journal from Antarctica, 1978–1979*, Ursula B. Marvin Papers, Smithsonian Institution Archives, SIA Acc. 19-001, Box 2, Folder 5, December 22.

Page 15: **"It is beautiful here . . ."**: Marvin, *Journal from Antarctica, 1981–1982*, SIA Acc. 19-001, Box 2, Folder 8, January 12.

Page 16: **"We are hiking . . ."**: attributed to Ursula B. Marvin; Marvin, *Journal from Antarctica, 1981–1982*, SIA Acc. 19-001, Box 2, Folder 8, December 28.

Page 23: **"It might not be as bad . . ."**: Marvin, *Journal from Antarctica, 1981–1982*, SIA Acc. 19-001, Box 2, Folder 8, January 14.

Page 24: **"This is the end . . ."**: Marvin, *Journal from Antarctica, 1981–1982*, SIA Acc. 19-001, Box 2, Folder 8, January 15.

Page 26: **"You've got a broken leg." "You are tough."**: Marvin, *Journal from Antarctica, 1981–1982*, SIA Acc. 19-001, Box 2, Folder 8, January 15.

Page 26: **"I am very sorry . . ."**: Marvin, *Journal from Antarctica, 1981–1982*, SIA Acc. 19-001, Box 2, Folder 8, January 19.

Page 32: **"I really would not exchange . . ."**: Kenneth Taylor, "Interview with Ursula Marvin," *International Commission on the History of Geological Sciences,* no. 34 (2002): 29.

Page 33: **"Riding through the vastness . . . a piece of rock from another planet!"**: Ursula B. Marvin, "Extraterrestrials Have Landed in Antarctica," *New Scientist* 97, no. 1349 (1983): 714.

SELECT BIBLIOGRAPHY

Fulton, Jacqui. "To Catch a Falling Star." September 21, 2018. In *Word of Mouth*, produced by New Hampshire Public Radio, podcast. https://nhpr.org/word-of-mouth/2018-09-21/to-catch-a-falling-star.

Marvin, Ursula B. "Extraterrestrials Have Landed in Antarctica," *New Scientist* 97, no. 1349 (March 17, 1983): 710–15.

Marvin, Ursula B. *Journals from Antarctica, 1978–1979, 1981–1982*, Ursula B. Marvin Papers, Smithsonian Institution Archives, Washington, DC.

Marvin, Ursula B. "A Meteorite from the Moon." In "Field and Laboratory Investigations of Meteorites from Victoria Land, Antarctica," edited by Ursula B. Marvin and Brian Mason. *Smithsonian Contributions to Earth Sciences* 26 (1984): 95–104. https://doi.org/10.5479/si.00810274.26.1.

Marvin, Ursula B. "The Origin and Early History of the U.S. Antarctic Search for Meteorites Program (ANSMET)." In *35 Seasons of U.S. Antarctic Meteorites (1976–2010): A Pictorial Guide to the Collection,* edited by Kevin Righter, Catherine M. Corrigan, Timothy J. McCoy, and Ralph P. Harvey, 1. Washington, DC: American Geophysical Union; Hoboken, NJ: John Wiley & Sons, 2015.

Marvin, Ursula B., and Glenn J. MacPherson, eds. "Field and Laboratory Investigations of Meteorites from Victoria Land and the Thiel Mountains Region, Antarctica, 1982–1983 and 1983–1984," *Smithsonian Contributions to Earth Sciences* 28 (1989): 1–146. https://doi.org/10.5479/si.00810274.28.1.

McNeill, Leila. "The Rock Star Geologist Who Mapped the Minerals of the Cosmos." *Smithsonian Magazine,* March 30, 2018. https://smithsonianmag.com/science-nature/smithsonians-rockstar-geologist-who-mapped-minerals-cosmos-180968640/.

Sandomir, Richard. "Ursula Marvin, Geologist of the Extraterrestrial, Dies at 96." *New York Times*, March 9, 2018. https://nytimes.com/2018/03/09/obituaries/ursula-marvin-geologist-of-the-extraterrestrial-dies-at-96.html.

Taylor, Kenneth L. "Interview with Ursula Marvin." *International Commission on the History of Geological Sciences (INHIGEO)* 34 (2002): 23.

Taylor, Kenneth L. "Ursula B. Marvin (1921–2018), Planetary Geologist and Historian of Geology." *Earth Sciences History* 37, no. 2 (January 1, 2018): 444–49.

Wood, John. "Ursula B. Marvin (1921–2018)." *Eos Magazine*, July 10, 2018. https://eos.org/articles/ursula-b-marvin-1921-2018.

SELECT TELEPHONE INTERVIEWS CONDUCTED BY SANDRA NEIL WALLACE

John Schutt, August 14, 2021
Dr. John Wood, August 22, 2021
Gayl Heinz, August 30, 2021
Tammy Peters, September 16, 2021
Harold Bailey, September 29, 2021